Weathering Agents

Carole Greenfield

Beltway
EDITIONS

Weathering Agents

Carole Greenfield

www.beltwayeditions.com

Printed in the United States of America 10 9 8 7 6 5 4 3 2 1

Book Design: Jorge Ureta Sandoval
Author and Cover Photos: Michael Greenfield
ISBN: 978-1-957372-02-0

Beltway Editions (www.beltwayeditions.com)
4810 Mercury Drive
Rockville, MD, 20853
Indran Amirthanayagam: Publisher
Sara Cahill Marron: Publisher

for the ones who've kept me steady
you know who you are
thank you

"The term **weathering** encompasses a variety of chemical, physical, and biological processes that act to break down rocks in place. The relative importance of different kinds of weathering processes is, in turn, largely determined by climate. Climate, topography, the composition of the bedrock or sediment on which the soil is formed, the nature of added organic matter, and time determine a soil's final composition."

—from *Physical Geology*, by Carla M. Montgomery

Acknowledgements

— "Certain Rooms Know Sunday Afternoons" first appeared in *Red Dancefloor*, 1991

— "In the Bakery" first appeared in *Women's Words: Resolution*, 1992

— "Ginkgo" first appeared in *The Sow's Ear*, 1996

— "Relative Dating" first appeared in *Arc*, July 2021

— "Rain Shadow" and "Luster" first appeared in *Beltway Poetry Quarterly*, March 2022

— "Braided Stream" first appeared in *Beltway Poetry Quarterly*, March 2022, under the title "Wrapped"

— "Upwelling" first appeared in *Sparks of Calliope*, June 2022

— "Eve" and "Gifts of Emptiness" first appeared in *Eunoia Review*, June 2022

— "Rill Erosion" first appeared in *Glacial Hills Review*, June 2022

— "Chilled Margin" first appeared in *Sky Island Journal*, July 2022

— "Outcroppings" first appeared in *redrosethorns magazine*, July 2022

— "Contour Interval" first appeared in *Autumn Sky Poetry Daily*, August 2022, under the title "Desire in the Time of Pandemic"

— "Before It's Not" first appeared in *Autumn Sky Poetry Daily*, August 2022

— "At Your Service" and "Disconformity" first appeared in *Dodging the Rain*, September 2022

— "Ground Truth" first appeared in *Dodging the Rain*, September 2022, under the title "Calling"

— "Daylight's Savings" first appeared in *Humana Obscura*, September 2022

— "Convergence" first appeared in *Sparks of Calliope*, September 2022

— "Foudroyant" first appeared in *Pulsebeat Poetry Journal*, September 2022

— "Shadow Zone" first appeared in *Solstice Literary Magazine* (now known as *Empyrean Literary Magazine*), September 2022
— "Spring Tides" first appeared in *Solstice Literary Magazine*, September 2022, under the title *Sometimes I Say Yes, Sometimes I Say No*
— "Hushed" won 2nd place in the Yeovil Literary Prize competition, September 2022
— "Blueberries" and "Crows" first appeared in *Amethyst Review*, September and October 2022, respectively
— "Ephemeral Streams" first appeared in *Vita Brevis Press Anthology*, October 2022
— "Trace Fossils" first appeared in *The Bluebird Word*, October 2022
— "Epiphany" first appeared in *The Plenitudes*, October 2022
— "Perturbed" and "Plucking" first appeared in *The Antigonish Review*, October 2022
— "Take Flight" first appeared in *Autumn Sky Poetry Daily*, November 2022
— "Lipstick Lives" first appeared in *Wizards in Space Literary Magazine*, December 2022
— "Locked Fault" first appeared in *Kairos Literary Magazine*, December 2022
— "Ice Cubes for Orchids" first appeared in *The Alchemy Spoon*, December 2022
— "Body Waves" first appeared in *Fresh Words Contemporary Poems Anthology 2022*, December 2022
— "Abrasion," "Cleavage" and "Nueé Ardente" first appeared in *The Quarter(ly)*, December 2022
— "Knot" first appeared in *Unfortunately Literary Magazine*, January 2023
— "Heat Flow," "Confining Pressure," and "Ring of Salt" first appeared in *Litbreak Magazine*, February 2023

— "Covalent Bonding" first appeared under the title "Hips" in *Litbreak Magazine*, February 2023

— "Hats Off" is forthcoming in *Ellipsis Literature & Art*, April 2023

— "First Footing" is forthcoming in *Floyd County Moonshine*.

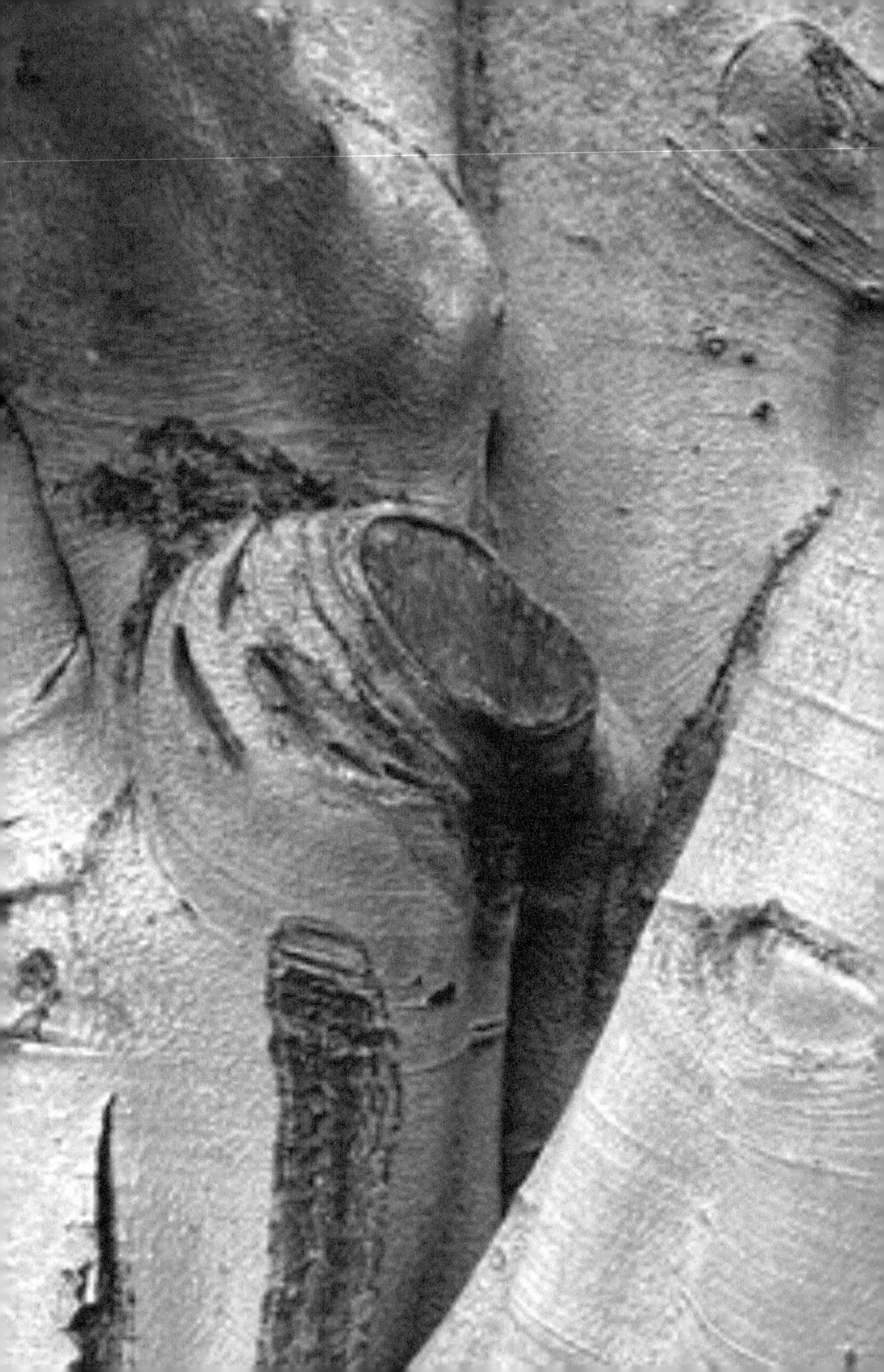

Table of Contents

Carole Greenfield

****

Weathering Agents

****

Part One

Luster

This time, it will last, the gift
of your smile, cello thrum
of your voice, smooth, sweet
linen of your skin.

This time, I will weave bright strands
of your touch into a silvered scroll: winter birds
and bamboo shoots, a span of mist-shrouded bridge
across an unseen river

and poems inked out in strokes swift, assured —
the way you moved within me as I closed my eyes
while light spread outward from my center,
coating us until we glistened in the darkness.

* Luster: *the surface sheen exhibited by a mineral*

Rill Erosion

The way I flow
in your direction, no matter
where or whether I am

your voice plays itself over
your touch carves
a path of light

fingers me lips to throat,
belly and beyond, unseating
my center, legs

shaking to point of sound,
clattering, chattering, battering
down my hatches, snatching

at sanity as hillsides crumble,
resistance tumbles, eroding
me to edged-raw

wanting that fills all
channels scooped out by your
tongue, takes and leaves me

wordless, clinging to the very forces
changing me moment by sloping,
shining moment

* Rill erosion: *soil erosion on sloping land by water forming very small channels*

Relative Dating

This is how it happened. What occurred and when,
and where. And how it changed the very air
we breathe, the eyes we see with and each
face, our sequence of events in place.

You moved close to help me off with my coat,
bent your head, kissed the side of my neck,
held your lips there 'til the breath caught
in my throat.

We knew then: there would be no retracing.
Light, unlacing the night, threaded us one
to the other as another morning sifted
through curtains while we drifted, half-sleeping,

half-waking, on darkened sheets. On the streets,
life clanked and rattled past. Inside, at the last
moment, we rose and left the bed.

Do your pillows ever miss my head?
Without our sweat, did the sheets forget
the way our bodies twisted, how your dark eyes
misted just before I came to understand
things would never be the same?

* Relative dating: *before any techniques for establishing numerical ages of rocks or geologic events were known, it was sometimes at least possible to place a sequence of events in the proper order*

Body Waves

Caught by tone and timbre, your voice's
shaping letters, cadence, words

I had to follow, place you
in my hemisphere, couldn't stop

to ask where we were headed
insinuating within crevices

put-to-bed desires, whispers
in my ear and I released

with a murmur, a breath
one by one you summoned

them to rise and swell
circle halfway round the earth

* Body waves: *the strength of seismic waves from a major earthquake is such that the body waves could, in principle, be expected to reach the opposite side of the earth*

Cleavage

Sound out my preferential zones
without so much as touching me

know how, know where
to place teeth and tongue

grasp and twist, hold and press
break me apart the way my legs

spread out my body and I take you
down, split all along our parallel planes

platy places where I was once aligned
where you mined your desire

bought and brought it
into line with mine

* Cleavage: *the tendency of a rock to break along parallel planes, corresponding to planes along which platy minerals are aligned*

Contour Interval

Something that I never did and now do every day
Count the hours that forever keep our lives apart
Tick them off my fingers one by one, each time I say
The only way that this will end is with a broken heart.

Count the hours that forever keep our lives apart
While our energy connection yet spells us and enchants
The only way that this will end is with a broken heart
Still we edge ever closer to the possibility of chance.

While our energy connection yet spells us and enchants
Our longing yearns across the continents and oceans
Still we edge ever closer to the possibility of chance
Reason is submerged beneath the storms of our emotions.

Our longing yearns across the continents and oceans
Before desire's rising tide, my defenses crumble
Reason is submerged beneath the storms of our emotions
All I want to do is take you into bed and tumble.

Before desire's rising tide, my defenses crumble
On soul and body level, we both know what we are missing
All I want to do is take you into bed and tumble
We shall spend those endless counted hours kissing.

On soul and body level, we both know what we are missing
Tick them off my fingers one by one, each time I say
We shall spend those endless counted hours kissing
Something that I never did and now do every day.

* Contour interval: *the difference in elevation between successive contour lines*

Shadow Zone

That first time, the force
of their meeting shocked body waves
to circle the globe; all peaceful beaches
vanished in the storm surge.

They shared sensibility with natural
disasters, mutual zones of weakness
and ferocity that alter landscapes forever.

Push down through tight spaces. Pry
free passion wrapped round withheld
wish, layered-deep desire.

Dig out while you still can.
So little warning once the waves
are on their way.

* Shadow zone: *an area of the Earth's surface where seismographs cannot detect P waves and/or S waves from an earthquake*

Alluvial Fan

Your hand caressed my head; I had to gasp for air.
That one time only; it hasn't happened since.
What I felt then – and now, cannot compare.

Of little bits and pieces I have had my share,
caught glimpses, received tantalizing hints.
Your hand caressed my head; I had to gasp for air.

Encountering the real thing, one cannot help but stare;
like seeing the original after only viewing prints.
What I felt then – and now, cannot compare.

That one time only, a memory cherished, rare,
became a tight-held secret, desire I can't evince.
Your hand caressed my head; I had to gasp for air.

Relinquishing sensation seems unfair;
exchanging jewel tones for pastel tints.
What I felt then – and now, cannot compare.

Would never having had it be the harder fate to bear?
One leaves you with a smile, one makes you wince.
Your hand caressed my head; I had to gasp for air.
What I felt then – and now, cannot compare.

* Alluvial fan: *a wedge-shaped sediment deposit left where a tributary flows into a more slowly flowing stream or where a mountain stream flows into a desert*

Nueé Ardente

Passion is not a comfort.
Be grateful to survive it once.

Twice in a lifetime is asking for trouble
or miracles, landscape transformed

until you become a stranger in your own body,
remembering the way her hair lifted off her neck

in a gold-colored breeze, taste of skin
where the hair had been

blood fireworks in the veins, lungs
thick with ashen shadows

breathe it in, that cloud-lit glow
desire and regret for what you left

or what you'll never know.

* Nuée ardente: *a deadly, denser-than-air mixture of hot gases and fine ash, from the French for "glowing cloud"*

Perturbed

The year I live alone a comet comes to pass. For a month,
I step barefoot into the street each evening to bid it goodnight.

So much simpler to locate its position near the horizon
than to think of you. That blurry ball of dirt and ice

shall leave and not return in this lifetime.
I never know if you are coming back.

Far off, you spark and beckon, slow glow
among stationary stars. Despite my wishes,

you will not stay. No spell to cast, neither
ancient rune nor incantation will halt your orbit

beyond the limits of my world.
Yet there is that part I cannot threaten

or cajole from out its stubborn faith. It keeps watch,
waits for you to gleam against the darkness,

slip inside and take me in your arms, stretch
your shining length across my body one more time,

leave a trail of dust and debris in your wake.

* Perturbed: *occasionally, a comet is perturbed in such a way that its orbit is deflected into the inner solar system, where it can be observed*

Convergence

I wish it were the other way round, evening hours (long stretching darkness into deeper darkness) yours and morning hours (black to gray to blue to gold) mine. Dawn has always been best, rising of my own accord (no need for clocks) to meet my grandmother at the pool, me swimming laps, she in her corner doing ballet, leaps, turns, legs like a girl's, smile dazzling as the sun flooding through floor-to-ceiling windows, drenching us both in light.

As long as I have known myself alive, I've loved the early morning, cycling down quiet sleeping streets to my job at the bakery, stocking trays, stirring oatmeal, salting grits, opening the door to customers lined up on the old porch, eager to enter, place orders, find a perfect chair and table, settle in for the best part of the day. Early hours. I can manage solitude in the morning. That time of day is never lonely. But late at night. Well. Quite a different realm. Separate hemisphere. Not my true home.

* Convergence: *a zone in which opposing water currents meet*

Dynamic Equilibrium

The smell, redolent; the flowers, pendant;
the vine, swift-growing, so strong Colette
wrote that it punched a hole right through
her window, wrenched free part of the iron
fencing. Believe it.

Terrible the pull, how passion
draws me back and back, invites
me to climb your trunk, sink
teeth, dig nails, wrap legs round
your solid bulk.

Tendrilled thoughts, images and dreams
wind and twine 'til I'm entangled, breath
by breath, smothered, going under, coming
up for air.

If not pruned twice a year, wisteria
grows beyond control. So far,
we've been cut back twice. So far,
no perceptible effect on taming
our rapture.

* Dynamic equilibrium: *the condition in which two opposing processes are in balance*

Chilled Margin

There is a great deal of the ascetic about me, you observed
after a night given over to the sensual. I continued to kneel
on the floor, arms wrapped round your waist, bones
of my cheek pressed against your ribs.

My heart beats strongly inside a thin chest, you told me
the first time. Listening to that quick thumping, I understood
it was the only thing sure and steady within you.

Nothing else could be counted upon.
Not your lips in all their beautiful precision.
Not your skull in my palms, the perfect bowl of it,
hard and smooth and infinitely precious.

Neither these nor your hands that stroked and fingered,
pulled and dug in rhythmic refashioning of my form.
All of them, those parts you shared so freely in darkness,
were to be taken back, retrieved without hesitation or regret.

You had cooled the way certain rocks do, rapidly,
a chilled margin like the stubble growing down your jaw
where you sat in my embrace, sealed by a bright glaze,

as if you had never rested warm and pulsing
in my arms, fitting the hard ridges and hollows
of your body to my deep curves,
my soft swelling.

* Chilled margin: *fine-grained rock at the margin of a pluton; shows the effect of rapid cooling*

Abrasion

Grind me down.
Nicks and cuts unending, accumulated like sand dunes,
like scabs I picked at in my teens, flaked edges 'til they bled,
left scars that faded but remained, pale indigo, unsteady
footsteps up and down my shins.

Wear me thin.
Slip a fingernail under an envelope's triangle to see
what's hiding within, how I turn over rocks and rotting
logs so my students can see life scurrying, burrowing
down in damp dark.

Suck me dry.
Leave a hollowed shaft of wood, good for nothing
but blowing through, then photograph withered
leaves, vines wrapped round their host,
loving it to death.

* Abrasion: *grinding erosion by rocks entrained in glacial ice or by windblown sand*

Plucking

Hammered dulcimers take time to tune.
Strings run in pairs, over one bridge,
under the other.

Plucked together along one side,
a pair sounds the same note.

When the hammers fall
on either side, this same couple, now a fifth
apart, chime in harmony.

Meeting as we did, only often enough
to check our hearts one against the other,
we kept ourselves in tune. A fifth apart,
we had our own concordance.

This is just a phase, you tell me.
A temporary rest in our tumultuous duet.
Your fingers pull back the strings;
I brace myself against the letting go.

* Plucking: *as water seeps into cracked rocks at the base of the glacier and freezes, rocks attach to the glacier. As the glacier moves on, it may tug apart the fractured rock and pluck away chunks of it*

Heat Flow

Press flannel-covered thighs against paint-flaked
corrugations, elderly radiator pulsing, hissing, unforced
steam clanking its way up and out.

Place clothes along scalded ridges, delaying that moment
of pointed pleasure, like draping towels dryer-fresh
over bare arms, remembered shock of hot
fabric's embrace (ancient comfort of heat).

Push naked legs against a husband's tensile thighs,
prehensile toes that pluck objects up with easy grace,
caress and press skin and bones (animal comfort of heat).

Pretend to forget what should stay trapped blue-deep
in ice, chip-edged, parts whose throb will not stay buried,
will beat and beat their way through, no matter what we do
to keep them in the dark, keep them cold.

* Heat flow: *the outward flow of heat from the earth's interior to the surface*

Spring Tides

Tell me you believe me when I say follow your dreams
Don't go out the window when you could open a door
Listen to the whispers or they turn themselves to screams

Step outside and feel me beneath the moonlight gleams
You say that we still have so much yet to explore
Tell me you believe me when I say follow your dreams

The story of our journey's written down our heartbreak seams
Try to pull away, box it up and lock the drawer
Listen to the whispers or they turn themselves to screams

The wise ones often tell us love alone is what redeems
Angels bring us back each time to even out the score
Tell me you believe me when I say follow your dreams

Not every part of passion's quite the glory that it seems
To gather broken pieces after crashing's such a chore
Listen to the whispers or they turn themselves to screams

Chart with me the river, merging of our two souls' streams
Why hide our desire or deny that we want more
Tell me you believe me when I say follow your dreams
Listen to the whispers or they turn themselves to screams

* Spring tides: *the most extreme tides, which occur when sun, moon and earth are aligned*

Rain Shadow

I like scraping lint off the mesh catcher, removing
/ still-warm towels,
sheets, bringing them to my face to breathe the heat,
/ faint sweet smell, folding them
well in thirds the way my ex-boyfriend taught me.
/ Fifteen years gone and his
influence still lingers, tingeing the way I go about my business.
How we continue to practice what the people we no longer live
with wanted when they loved us, when we loved them.
/ I still place knives edge up
the way my first husband did, still does for all I know.
/ Second one taught
me to put things in the sun to cure, bleach bad smells,
/ drive away all signs
of sin. Maybe I should put myself there for an entire day, burn
away all impurities, leave behind nothing save a dried-up husk,
all my juicy, pulpy, seedy, crunchy sweetness smothered out.

* Rain shadow: *as moisture-laden air from over the ocean moves inland across the mountains, it is forced to higher altitudes, where the temperatures are colder and the air thinner. Under these conditions, much of the moisture originally in the air is forced out as precipitation, and the air is much drier when it moves farther inland and down out of the mountains. In effect, the mountains cast a rain shadow on the land beyond*

Ground Truth

All I can see is your wedding ring, you say when I try to show
off my favorite panties, donned in your honor, minky pink faux
leopard print, about as wild as I get, unlike you, resplendent
in crimson briefs you washed with the t-shirt you slept in last night,
pale rose, gentle as the insides of a conch, the kind raised
/ to lips and blown
to call our faithful to prayer, the way you answer the call
/ each morning,
walking to your congregation where you pray for me, you say,
each day you pray for me but today asked me to do so for you.

Will these be the last words I hear from your lips? I shape mine
into a kiss, blow out my breath to release the harsh gripping
/ and recall
a sudden vision I had while watching your face: sitting beside you
in your garden, dappled shadows touching us as I reach to stroke
the white hairs by your ears, pass my fingers over them so lightly
you can't feel me but still you know I'm there, answering your
/ call to prayer.

* Ground truth: *check of the interpretation of remotely sensed date by direct contact (observation on the ground)*

Confining Pressure

Deep within barren rock, your truths lie
buried, striated strands of dimming light

Layer upon pressured layer, innumerable nights
together, brushed desert pastels, sunrise hues

That life to which you belong enfolds,
holds you against wavering lines of stone

Edges bleached and blurred to comfort
delude, do not deceive, cannot be breached

Ambient patterns of your choosing
you will never choose to whittle free

* Confining pressure: a*mbient pressure equal in all directions, imposed by the surrounding rocks, all of which are under pressure from other rocks above*

Disconformity

Terracotta, shades of slate. Looking out on winter's end
while you walk among sun-bricked remnants of civilizations lost,

I watch future ones run to line up on asphalt-marked x's showing
where to keep distance. Your rocks tumble over each other;

my young charges dash past, twirl, twist in shapes I couldn't get
into if I tried. You ponder beds of broken prisms, polygons

my little ones learn to name at age five. I teach other people's
children, will never carry any of my own. You wear the most

recent addition strapped to your chest which I will never press
against the way I lie upon my husband's in late afternoon light

watching shadow bars on the wall as he slips into sleep and I try
not to think about you. Are we broken inside, lying at proper angles

in rock beds of our own making? What right have we to wish
for contours that might fit the people we eventually become?

* Disconformity: *an unconformity at which the bedding of rocks above and below are parallel*

Locked Fault

Locked in our faults, sacred vaults, spaces
created from manifold connections, intersections
keeping chains firmly fastened round wrist and ankle,
heart and tongue.

No matter how long we lock legs, fingers,
lips, nor how neatly we slip between,
there will come that moment when...

so let's not speak of fault, my dear, yours
or mine, theirs or ours, hers or his, but do
consider this: in your life, what am I displacing?

In mine, what are you misplacing?
And can any of it be replaced
once they've changed the locks?

* Locked fault: an *active fault where friction is preventing stress release through creep, no displacement is occurring*

Ventifact

Stone archway over flagstones laid in earlier epochs, sun-drenched,
sun-baked, longstanding, undemanding, understanding it will outlast,
outlive all beings who pass under, press palms against, run fingertips
along scarred surfaces impenetrable, imperturbable, implacable.

How many gateways will we go through in one lifetime?
Of all the doors that close behind us, which can we go through both
ways and repeatedly? How often do we miss the mark, think
we'll find our way back before we learn the latch has locked
behind us and there is no key for the finding?

Some gates we have to close ourselves, others shut without
/ our say-so,
whether we will it or no. For our own good, we are told.
For the good of the world. Repairing the world.
At the cost of our souls? What a cost.

* Ventifact: *a rock sculptured by wind abrasion*

Braided Stream

Brittle-stringed leaves wrapping round, around, slow revolve
/ on your turntable.
So many loving names we had in the time of our entanglement,
our lives our blood our breath entwining as limbs of lovers do,
/ the way
ours have not yet, perhaps never will, the way my husband
/ wraps his tense-muscled
legs over and under mine, toes reaching for my chilled feet each
time I come back to bed. We go deep into one another. Our roots
push down. Your deft-fingered words tie me into new knots,
/ enwrap my heart,
enrapture and undo me. The time for unwinding is not here,
/ is not yet, is not now.

* Braided stream: *a stream with multiple channels that divide and rejoin*

Continental Drift

They were always struck by how we fit, so well
the prospect of separation grew unthinkable
as the idea of continents unlatching.

A few fissures here and there, hairline
fractures, requests denied, wishes postponed,
gradual gathering-in of resentment, then resolve.

Point of no return. Salt sifted
into flour. No route back.

Study the Pacific coastlines of Africa
and South America, see how easy to believe
they once meshed.

The way Brazil reaches out toward Gabon,
my heart still yearns in your direction;
how I live with the aching of spaces

you once filled so beautifully
I never imagined a time when your life
would travel in directions apart from mine.

* Continental drift: *the idea that individual continents could shift position on the globe*

Trace Fossils

Small children do not wait for pain
to make a lasting mark. They give fair warning;
we have time to wipe off tears, mop up trouble, kiss
a bruise, pronounce it healed.

But love leaves an impression that won't
be kissed away. An imprint left in something soft
hardens and congeals. What passed through fire once
is tempered, then annealed.

Children trace fingers over fossils, guess
at what's revealed: evidence of ridges, indentations,
life long over, heart's rush sealed.

* Trace fossils: *fossils in which evidence of organisms, rather than the organisms themselves, are preserved*

Upwelling

Last hour, last day before break, alone in an empty classroom,
/ is when
I hear them, sleigh bells, silver jangling in the air, then steady thudding.
Before I can leave my chair, a line of children gallop by, shaking
handbells. By the time I reach the doorway, they are past me,
/ heading down
the corridor. No one sees me watching, not even the child at the
end, my Alejandro. Alejandro, who tries to speak English, gets
embarrassed, says, "Forget it," folds his arms and closes up his face.
Alejandro, whose sudden smiles I can count on one hand and
/ still have
a finger or two left to hope on. Alejandro, a skeptic at
seven, if not yet a confirmed cynic. This same person brandishes
his bell on high, proclaims to the world, "I like!" and leaps
/ twice in the air,
rounds the corner, vanishes from view. I stare at the space that
/ held him,
bells, echoes of clear joy resounding, reminding me of secret hearts
that try to reach the surface no matter how much weight we
/ make them bear.

* Upwelling: *the rising of deep, cold waters to shallower depths in response to reduced surface pressures*

Outcroppings

It's slate if it draws on rock, I tell him. He is not the easiest.
I seek ways beyond the pale, outside the lines. We hunt for rocks,
quartz and mica, slate and shale. 'How much do you think it's worth?'
he demands, thrusting a sample close to my face.
'A hundred dollars? Two hundred? A million billion?'
Remember what we read? I say. *If it's easy to find, it's not worth much.*

It's the way of the world. We value only what we lack or possess
in reduced amount, failing to appreciate what's here: a husband
who listens, forgives your gloomy spells and minor transgressions;
ease of motion until a foot cramps, a nerve twangs; ignoring
songs of sparrows and starlings, loving only cardinal and wren.
The everyday joy that tints our daily details, we overlook;
make time to note them down. We must.

It's slate if it draws on rock. He scratches a dark line
along the sidewalk, shakes his head, tosses the shard away.
I think of petroglyphs, thin-lined figures
dancing on rock, scratched out by other human hands
seeking a way to note the ordinary
extraordinary love that lifted hearts.
It's about being alive while we can, dancing through
our present, leaving our mark, not knowing how or if, or when
it may be read, centuries from now.

* Outcropping: *a coming out of bedrock or of an unconsolidated deposit to the surface of the ground*

Covalent Bonding

At twelve you already knew why you were an atheist.
I was certain of nothing but that I'd gone
months without a boy for a friend and wanted it back,

the uncomplicated mystery when holding hands
was just a lovely place to put your fingers.

The year you were twelve and I one year older,
nothing was easier or more joyful than lying
beside you in the field behind our houses,

warmth on closed eyelids,
clouds silent and slow,

and at night there was walking your dog,
her pale hips lighting our way.

* Note: *The chemist Kekulé once envisioned covalent bonding among carbon atoms as a kind of hand-holding among atoms*

Ephemeral Streams

Just back from two years in Africa, you had pictures
in your hands and in your eyes. I looked through desert places,
one where you had lived, one that lived in you, parched,
the two of the them, holding memory and hope of rain.

You had stories gathered up the way those girls used
fingers to string fish through gills. I wanted to reach in,
pull them out, one at a time, slowly. Taste them
in my mouth.

I understand something about dry places,
the way edges harshen and color bleaches out,
how one sees clearly for a long way off,
when it gets harder to hide.

But I was in love with rain
and the change it wreaks on landscape.
How things dissolve and wash away.

You saw the end approaching from the moment
we began. I did not argue. One doesn't
with weather.

When the time came for you to go,
I watched you leave, grown beautiful
and strengthened by all that had flowed
between us.

* Ephemeral streams: *streams in deserts, which may flow only during a short rainy season or even just for a matter of hours after a cloudburst*

Terminal Moraine

The world I loved is gone, and I am out of place
I mourn a time when people had to wait
a world where you could still talk face to face

I seek a world that's left without a trace
the busy signal, letters that came late
The world I loved is gone, and I am out of place

When libraries held only books, I could keep pace
with what there was to learn; I knew my fate
among the people who talked face to face

I've tried; I can't compete in today's race
The kind of speed required not a trait
I love. I would that I were gone from out this place

The changes that have come I can't erase
What matters to me used to carry weight
when people looked you in the eye, they saw your face

You'll tell me I am just a hopeless case
I know. All that remains is trying not to hate
this world in which I feel so out of place
Mourn what is lost, and face the truth we have to face

* Terminal moraine: *the end moraine marking the farthest advance of a glacier*

Part Two

Eve

Ascend the rungs of your ribcage,
God instructed us when we first learned
to breathe.

I try to climb slowly as panic gathers
but my breath snags on that rib
of his.

I don't believe I'll ever reach
my collarbones.

Do you think of leaving?
he murmurs, holding me on his lap, his lips
warm along my jaw. I say nothing, shake
my head no. Where on earth would
I go?

At night while he sleeps beside me, mouth open,
I imagine removing that slender bone
beneath my breast, boring holes
at either end,

five more down one side,
putting it to my lips and walking out
past hedges, briar roses,
and the last tangle
of wisteria.

Not even God could name
this new sound,

the melody of lungs filling and falling,
only my own ribs around them.

Hushed

I who wash the dishes after every meal,
never leave so much as a spoon unrinsed,
set the glasses open ends down and draining,

I who stow leftovers and replace perishables
in their proper bins, sweep the floor
and wipe the counters free of crumbs and stains –

I left it all:

plates on the table,
pots on the stove,
food on the counter, unwrapped and warming

when he put his arms round me and found
my mouth with his. I let him
walk me step by step

into the next room, lay me down
as one would a precious garment
spread out and admired

as he covered my face
with his hand, fingers
light against my forehead

and cheekbones, palm
over my nose and lips.
"Hush," he said.

I breathed in slowly and with my tongue
found his love-line, felt it
run its course.

First Footing

I loved my first husband's feet, his beautiful legs
that bring to mind a long, quiet hall filled with light,
white statues of young Greek men and their gods.
I walk slowly round them, trying not to look

as if I'm looking. I want to run hands
up and down their cool, muscled stone,
work fingers into each and every crevice.

I remember the feel of his thighs, shape
of his feet, casual grace and firm-planted strength.
Those legs and feet root him wherever he goes.

In loving other men's feet, I engage
in a fruitless search for the one pair I chose
to walk away from. Like Lot's wife,
I looked back, am still gazing

over my shoulder – hair whipping my cheeks,
ends stinging my fixed-open eyes –
at a life I can never reclaim.

Ring of Salt

It's just chicken with a ring of salt, she wrote, later sent
an image of a carcass surrounded by a mountain range

of silvery grains and I thought, *But why?* What earthly
purpose would it serve other than to make food unbearable

and ward off all beings come to wreak havoc or spread light?
Why else deposit salt mounds round? Guardian angels

or warden officials, the way the Andes Mountains
ringed the city where I grew up. No one passed

the threshold of their eternal vigil.
I remained intact.

Knot

A covenant, we called it. Something to enter.
I stepped over that threshold as I would edge
my way onto an unfamiliar craft, noting its angle
of flags.

He knew all about the tying of knots.
Anchor. Figure-of-eight. Inside clinch.

I worked them loose and formed another kind,
ones I needed for my journey.

Half hitch. Throat seizing. Thief.

Some promises slip free no matter how tight the fastens,
secure the hold.

Daylight's Savings

Light tinged with trees' fast-fading gold,
wearing off, rubbing thin, last gleams caught

in the end of an afternoon I stand by the window watching,
knowing it is one hour later than the clocks say

knowing I will never see your eyes without
recalling that first time they shone out at me

unguarded in the swift-edged twilight, the early
darkness drawing down.

Crows

Raucous ballet of dark birds, cries sawing at the cold air,
flapped in staggered sequence, landing of one a cue for the next

to take heavy flight in brief spaces between branches, feathers
shifting ebony to chrome, chorus of tarnished angels overhead,

miracle of somber, hoarse-voiced beauty, plaintive
threnody stinging me to tears as I turned to see you

also watching, elbows folded on your car's roof,
gaze lifted to those gold-and-silver birds.

Not every love is as you'd pictured. Not every gift
comes wrapped and labelled with your name.

Blueberries

This morning I ate blueberries, tart and sweet
slight grit, soft explosions in my mouth.

I thought of my nephew, recently introduced
to the pleasures of eating, now able to feed himself in fistfuls

with those long fingers he folded beneath his chin
the day he was born; I saw the Buddha in his hands.

He seems to hold the secret of contentment in the world
he inhabits; almost everything is cause for smiling,

is reason to reach, point, look, explore with face
and mouth and skin.

When he eats blueberries, it is enough to fill his sweet body
with such satisfaction his whole being radiates delight.

He stretches a finger in my direction and I bend closer
to receive his benediction, the blessing of his sticky touch.

At Your Service

I was the one who got up at dawn to cycle down roads empty
and warm. Who opened the office, counted the till, set up
the drawer. Tore apart biscuits, piled pastries on trays to look
tempting, unlocked the door to let you all in. I was the one
who frothed milk for your lattes, spooned oatmeal and grits,
with cheese and without. Who told you the cost of your blueberry
muffin, your bear claw, your scone. Wiped down wooden tables,
stacked china plates, hung dish towels to dry. I was the one who
avoided the questions, *"And what are you doing? Working where?
Doing what?"* Who ducked disappointment of parents and friends.
Never told them I waited on angels, regular patrons I greeted
/ by name.

I'm still on that road that I started of serving, observing, gathering
up stories and fragments to serve me for years.

Certain Rooms Know Sunday Afternoons

In the room where no one waits,
bamboo blinds sieve almost-dusk.
Concave to the window, they tremble occasionally;
indoor weathervanes.

A couple on a mattress,
fingertips on eyelids, lips pressed next
to shoulder blades – their bodies hiss
across the sheets. She holds back
her hair with one hand, smiles; his palm
cups the space around her skull, fingers shaking.

Late-day winter never gleams. It seeps,
filling the room with a silence the color of clouds.
Boundaries loosen, landmarks fade.
She knows this as she looks up after
to see blinds sway near panes
locked against the cold.

Foudroyant

the quiet stun
of your touch, dark heat
of your smile, its private sun

in the last hours of blackness
before bird calls, gray beyond glass,
footsteps in halls

outside your room and on the street,
other feet, the weight of rain
marking pavement with sound
with stain

still shadows wet and glistening,
my heart is quiet, listening,

where I was struck,
I gleam

Hats Off

It's the face that makes the hat, my most handsome student told
/ me, decades
ago, back in the day when hats were my signature fashion statement,
one way to set myself apart. *You can carry it off*, my mother
would say and add, inevitably, *I never could.* Hats of all kinds
I had, much like my mother's father, collector of yarmulkes, silks
and satins, velvets, suedes, stacked in a desk cabinet we'd open first
thing each time we came to visit the small maroon house
/ tucked beneath
maples whose seedpods we'd split, affix to our noses, strut
/ round the yard,
making our grandparents and mother shake their heads at us
/and laugh.

The yarmulkes we piled up on our little heads, paraded through
living room, dining room, kitchen, calling out, *"Caps for sale!*
/ Caps for sale!"
just like the old man in the story, shaking his finger, then his fist
at the monkeys. He never shook his fist at us, our grandpa, although
he had a temper, but by the time we knew him, he'd become a mild
old man who cried at first sight of our mother, who held him
/ in her arms,
saying over and over, *"It's all right, Daddy, it's all right"*, and tried
to explain to her children that there was nothing wrong, he was just
happy to see her after so many long months each year
/ we were gone.

Ice Cubes for Orchids

Even when I've lost track of days, I know
when it's Wednesday, the day I give ice cubes
to my orchid.

It's not that I'm a fan. I'm not. True,
some call them beautiful. Exotic, erotic,
delicate, temperamental. Like myself,
except I'm neither exotic nor delicate, though I can

be quite temperamental. Just ask my husband,
that sweet, long-suffering man who gave
me this small plant that sits in its translucent
container by a window not in direct light.

It needs diffuse lighting; like me, can't take
full-on sun, never could, always preferred
gray light, the half light, rain and the mist light.

He gave me this orchid plant for Valentine's Day
more than four years ago, when I hadn't the heart
to confess I've never liked them. He's never liked
ice cubes, prefers his water at room temperature.

An orchid needs ice cubes one day per week.
Each Wednesday, i crack out
two cubes from their blue plastic tray, wedge
one on either side of the dark green leaves,

softly encourage the little plant, so stubborn
in its faith, much like my husband, gentle,
quiet and unyielding in devotion
to his wife, work, family and friends,

to plants in our garden, no matter how ragged
or withered, how on the verge of perishing
we might be; as long as one leaf lingers alive, he will
continue to tend, to care, to love.

Take Flight

Late summer sun sifts through a screen of trees
Still dripping from my swim, I see her taking flight
Completely naked, tiny feet tucked up to knees
Her fingers with her father's locked in tight.

If we could only return to the place
where we were all their world, our parents' goal
to keep us safe from harm, and one day face
our restless lives with hearts intact and whole.

The scene plays out before my tear-wet eyes
A joyful spirit swung into the air, delighted cries
And as her father steps away from shore,
she weeps, "Again. Again. I want some more."

He gently calms her tears. "No need to cry.
Just tell me. Say the words, 'I want to fly'."

Epiphany

And what did they come bringing you this year?
And what would you have wished for,
had you even been asked?

And what would be the proper response?
'Nothing, thanks.' 'I've everything I need, thank you kindly.'
'Thank you for asking; couldn't be better.'

My young student when I posed the same question
stood up on a chair to show me how long
she'd wish her hair to be

beyond her little bottom, to mid-thigh

the very place my gray chemise
grazed my legs the first time
I danced for you

morning light pouring through
kitchen windows

my inner sun at last matched
by the light of day.

Lipstick Lives

I want to be a Tuxedo Red woman:
stride into a city and take it by storm,
descend without warning, catch you off guard.
whirl you away, leave you gasping for air.

I want to be a Champagne Gold woman:
waltz into a party and take it by surprise,
sunshower when you expected fair weather,
water in the air like sequins rippling down
the front of a gown, flash of skin snatching
your breath and flinging it back with an edge
of gleam, hope and thoughts of parts unseen.

I want to be a Shadow Rose woman:
slip into a heart, take it by degrees,
a rain you wake to find dripping from the eaves,
leaves of trees marked out in the dark glow
of a wet spring morning,

and how your thoughts are forming
to fit around the stranger who smiles,
stakes a claim and stands
her ground.

Before It's Not

Today I've had the opportunity to watch bees
burrow in and out of pale-green hydrangea blooms,
listen to breezes in neighbor oak and maple,
as well as hammer and whine of buzz saws
on a nearby addition going up in place
of two lovely tall trees taken down at summer's
start. I didn't realize how much morning
shade they gave 'til they were gone.

I try to notice what I have, what is here
before it's not. I try to give thanks, practice
humility, manifest appreciation

for the giant oak tree still standing on the corner,
for what's left of the maple tree behind
my next-door neighbor's garage,

for each day I waken in a body free
from pain, for each day my parents are yet living,
each day I waken to the world.

This I know: bad things that happen
are never ones I dread, but what I never
thought possible.

My husband swung the spade so hard into my finger,
it left a blood blister that lasted for weeks,
a dark splotch, tender to the touch.
Could have been worse. Could have been
the whole finger whacked off.

This I know too: a day with low humidity
in New England is a good day, even though we need
the rain, desperately; it's still a beautiful morning,

and being able to wake up, brew tea,
step outside to see a goldfinch
perched on a drought-browned echinacea,

digging out seeds for its breakfast, well,
that's enough for now. Maybe even
a little bit more.

In Accord

You were once my lifeline; what connects us now is just as real
As the stretchy cord that spun me out and wound me back
Stepping over and around it, you prepared the evening meal
Kept our home together, our family life on track

As the stretchy cord spun me out and wound me back
We tried to keep from getting in the other's way
Kept our home together, our family life on track
Listening to my calls was how how you learned about my day

We tried to keep from getting in the other's way
The cord stretched taut between us, gentle tug, a little twist
Listening to my calls was how you learned about my day
I moved about the kitchen, cord wrapped tightly round my wrist

The cord stretched taut between us, gentle tug, a little twist
Attached to the wall as I was once attached to you
I moved about the kitchen, cord wrapped tightly round my wrist
And then I had to break it. It's what our daughters do.

Attached to the wall as I was once attached to you
The cord between us cut but not unbroken
And then I had to break it. It's what our daughters do.
The love between us sometimes went unspoken.

The cord between us cut but not unbroken
Stepping over and around it, you prepared the evening meal
The love between us sometimes went unspoken
You were once my lifeline; what connects us now is just as real.

Gifts of Emptiness

Two empty glass honey jars on my kitchen sill, second wedding
gift from the woman who guided and inspired, arranged my bridal
blessings shower, left our world too soon. Her last words
to me, *"Dreaming of spring and your exuberant gardens!"*
I look at her jars as gifts of emptiness and abundance, gathering air,
intangible fragile strings of nothing from which to weave a
/ tapestry of loving.
I look at my echinacea and false indigo, wind anemone,
/ bleeding heart,
butterfly bush, larkspur, tell myself she's coming back,
/ she's somewhere
to be found, an unfinished lyric or line of song, the notes are
/ there, just beyond
the page, just past my fingers. I can hear her voice, I can feel
/ his touch,
I can smell the perfume my mother wore, the tobacco
/ my father smoked,
I can see the woman I used to be, if I hold the empty jars and
/ close my eyes.

Gingko

He could lose all his leaves
loving her

present to the world
bare-headed, spare-boned

wind touching him
everywhere

rain and sunlight finding
all his spaces

He would need to put his secrets
in deeper holds

or let them show
let them go

In the Bakery

Sienna-skinned, she waits on us,
patient behind glass, silver trays
balancing cakes, pies, a fragrancy of cinnamon
spiking the close air.
Behind her, hollows in the wall,
dug-out shelves of adobe
painted white and dark with bread,
loaf on loaf stacked above her head,
her hair the nut-brown of crust.

My father is troubled by verbs.
He points at what he wants,
a crumble-topped cake shaped like a color wheel
shading buttercup to maple. It's his favorite.
She knows from months of Sundays,
and smiles, wrapping.

'Anything more?' she asks politely, her syllables
slowed for him. He nods.
'Two of bread,' he answers,
and I love him for his firm awkwardness.
She twists her body carefully,
as if she were trying to protect it,
and searches the depths of the loaf-homes,
shelf on shelf of variegated breads, wheat and rye
and other grains whose names I haven't tasted.
'Light or dark?' she says to my father, but her words
sift past the mesh of his limited vocabulary
and he stands on the floor, helpless and smiling,
clutching his cake.

Before I can lean up and whisper,
she strains across the counter.
The smells of yeast and sugar seep out from her creases
as her fingers touch my hair.
'Like this?' and then, retreating, taps her own crown,
shining walnut in the dim interior. 'Or like this.'
My father's heavy eyelids lift. He stares at the woman
whose face is lighted in reds and browns,
covers my head with a weight that cups
my skull and soothes, and smooths.
'Like this,' he tells us. 'Like this.'

Carole Greenfield

grew up in Colombia and lives in New England, where she teaches multilingual learners in a public elementary school. In the previous century, her work appeared in *Red Dancefloor, Gulfstream and The Sow's Ear.* More recently, her poetry can be found in such places as B*eltway Poetry Quarterly*, *Amethyst Review*, *The Plenitudes* and *Dodging the Rain*, among others.

WEATHERING AGENTS

PRINTING WAS COMPLETED IN MARCH 2023 FOR **Beltway Editions**